THE LITTLE BOOK
OF MEDICINAL PLANTS

THE LITTLE BOOK
OF MEDICINAL PLANTS

Élisabeth Trotignon

CONTENTS

LA BRIONE

THE HISTORY OF MEDICINAL PLANTS

Since the dawn of time, humans have been interested in plants, both as food and medicine. People gradually came to know about them through trial and error. They were guided by their instincts and experiences, which could sometimes lead to failure, or worse, death. But thanks to their keen sense of observation, they have built up a reliable knowledge. They attached importance to the virtues of plants, sometimes seeing them as the source of supernatural forces.

From popular wisdom ...
In the past, country dwellers had their recipes. They were extremely familiar with the land, and when the time was right, they would go out to pick plants for making herb teas and ointments. They knew that trees and flowers had the ability to treat certain ailments, which would later be confirmed by scientific research. They also trusted in the healing power of their parish priest, in old men reputed to have certain magical powers or in healers, who would supply them with secret remedies made from plants. But a cure was never guaranteed. Botanical wisdom became increasingly reliable and was passed on from generation to generation. Even today, there are still people who continue to treat themselves with the natural remedies they inherited from their ancestors. Unfortunately, modern societies have almost

L'Œillette

completely forgotten the wisdom of the past, relying now only on synthetic molecules.

... to scientific knowledge

In ancient times, the Chinese, Indians, Egyptians and Romans used plant-based remedies for their ailments. But it was the Greeks who did the most to advance botanical knowledge. In the fifth century BC, Hippocrates, the father of medicine, was already recommending the use of white willow leaves to reduce fever. Two centuries later, the philosopher and naturalist Theophrastus described the medicinal properties of more than five hundred plant species. In the first century AD, it was Dioscorides' turn to write his famous treatise *De Materia Medica*. Afterwards, the botanical knowledge established by the Greeks was constantly taken up and repeated. In the Middle Ages, Arab scholars and physicians as well as Christian monks and nuns copied the ancient manuscripts. They also studied the plants recommended by the *Capitulare de Villis* (a document from Charlemagne's time containing a list of flowers, trees and herbs, among others, to be grown throughout the empire) and by the physicians of the school in Salerno, and planted them in their gardens to study their benefits and dangers. Among the most famous botanists, Hildegard of Bingen in the twelfth century and Albertus Magnus in the thirteenth had thoroughly studied their predecessors' writings and contributed their own work to the advancement of botanical knowledge. During the Renaissance, the Italian Pietro Andrea Mattioli distinguished the species of his country from foreign varieties. In Germany, two monks, Hieronymus Bock and Leonhart Fuchs, known as the 'German founders of botany', created a number of herbaria and expanded knowledge of medicinal plants. Elsewhere in Europe, Charles de l'Écluse and Matthias de L'Obel of Flanders and Conrad Gesner of Switzerland were doing the same. It was also the time when Paracelsus developed the 'Doctrine of Signatures', according to which any ailments affecting parts of the body had its remedy in plants resembling the affected part. For instance, yellow flowers such as the marigold were meant to be used to treat jaundice. In this revival of the botanical sciences, travellers played a leading role, bringing back from distant lands new medicinal plants such as ginger and senna, and later Peruvian bark and eucalyptus. In the wake of the great botanists of the seven-teenth and eighteenth centuries, Joseph Pitton de Tournefort, Linnaeus and the Jussieu brothers, among others, came the chemists and pharmacists of the nineteenth century who studied active ingredients derived from plants. To them we owe the discovery of digitalis in foxglove, morphine in the opium poppy and quinine in Peruvian bark. Today, more than 40 per cent of synthetic drugs are made from copies of the natural molecules found in leaves, bark and roots. This practice does not detract from the merit of medicinal plants, which harbour great potential for healing. Furthermore, with trust in chemistry coming into question nowadays, the use of medicinal plants is reassuring.

WARNING

The information provided in this book reflects commonly accepted knowledge at the time of its writing. This book is not intended to be a guide to medicinal plants. The antique illustrations contain certain errors and lack accuracy. On no account are they to be used as models. The methods used to identify plants are very complex. It is therefore essential that any plants picked in the wild or in a garden are checked by a doctor or pharmacist. Likewise, regardless of the intended internal or external nature of their use, it is always advisable to consult a book by an expert explaining the precise dosage and instructions for use for a plant. Certain plants pose a risk of poisoning and deserve particular precautions, which are explained in this book.

YARROW
Achillea millefolium

This plant's scientific name is taken from Achilles, the famous hero of the Trojan Wars, who is said to have used it to treat his wounded heel. But this story also appears to be as much a legend as that of the Greek warrior because yarrow, which abounds in the central and northern regions of Europe, does not grow in Greece. The leaves are cut into very thin strips, releasing a strong, camphor-like aroma. The tops of the plant flower with clusters of white, sometimes pink, flowers at the height of summer. They are mainly used for their healing properties, hence the many names by which this plant is known, such as woundwort, stanchweed and carpenter's weed. It is used externally to treat skin ulcers, chapped and cracked skin, impetigo and bruises. In the past, it was also made into a decoction and used to treat scabies in sheep. The flowering tops aid digestion, and if taken as an infusion before meals, stimulate the appetite. As a decoction, they stop nosebleeds and relieve haemorrhoids. When used in a footbath, they can eliminate foot odour. While yarrow has all the makings of a miracle plant, it can also make skin sensitive to light. Therefore, the treated area of skin should be protected from light during use. This plant can also produce allergic reactions, and it is not recommended for use by pregnant women.

LES PLANTES UTILES

MILLEFEUILLE

ALOE VERA
Aloe barbadensis

This robust perennial plant, which was known to the ancient Egyptians, is native to eastern Africa. It spread to the warm Mediterranean regions, where it was propagated by simple cuttings. Today, it also abounds in the warm areas of the Americas, where it was introduced by the Spanish. This succulent plant is identified by its large fleshy leaves and orange flowers clustered about a long, stiff spike. A yellow, bitter concentrated juice is extracted from its leaves, together with a clear, viscous gel, sold under the name of aloe vera. Depending on the daily dose taken, the juice can be used as an appetite stimulant or laxative. The gel, in turn, has been known since ancient times for its remarkable skin-healing properties. When used in a compress, aloe vera cleans wounds and burns and speeds up the healing process. A historical account shows that Alexander the Great used it to heal an arrow wound that had become infected. In lotion form, aloe vera soothes skin irritations. Today, aloe vera is in demand by the cosmetic industry, where it is used as a moisturizing ingredient in make-up removers, lipsticks and sun creams. It is also used to make anti-ageing skin creams.

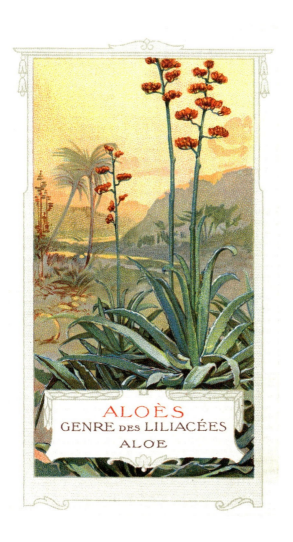

ALOÈS
GENRE DES LILIACÉES
ALOE

ALMOND TREE
Prunus amygdalus

Native to western Asia, the almond tree spread to the lands around the Mediterranean Sea and to China in ancient times. It is able to withstand intense sun and high temperatures, and it grows in arid and rocky soils. Its white or pink flowers appear before the leaves, and it can produce both sweet and bitter almonds. While *Prunus amygdalus* var. *dulcis* produces the former, which are edible, *P. amygdalus* var. *amara* produces the latter, highly toxic fruit. Unfortunately, it is impossible to distinguish between the two species growing in the wild. For safety's sake, it is better to buy almonds in a shop. Unlike the bitter almond, the sweet almond is very beneficial to humans. It is nourishing because of its high content in the B-complex vitamins and vitamin E, phosphorous, magnesium and other minerals. Diabetics can eat almonds without fear because it is low in carbohydrates. When used as a cosmetic, almond oil – provided it is freshly extracted – soothes and softens the skin. It also relieves itching. When used internally, it relieves inflammations of the digestive tract and constipation. It is also an excellent table oil for accompanying fish and poultry, although it has the flaw of turning rancid very quickly. Almond milk and orgeat syrup – a syrup made from almonds, sugar and water – are highly refreshing.

AMANDIER

ANGELICA
Angelica archangelica

The angelic reference in the name of this plant refers to the belief that the archangel Gabriel told of its powers as a remedy against the plague in the Middle Ages. This large aromatic plant even grows in Scandinavia, where it is found in gardens. Although the physicians of antiquity were not familiar with it, it was considered to have magical properties by the Middle Ages, when, worn as part of an amulet, it was used to ward off curses or bewitchment. Angelica has many qualities, which explains why it is often found in gardens. The leaves and stalks can be candied for use in confectionery or pastries, and the plant can be used to make liqueurs, such as Chartreuse and Bénédictine, the secret recipes that were jealously guarded by the monks who made them. But angelica was mainly sought for its proven medicinal properties. In an infusion or decoction, the roots, tender leaves and seeds stimulate the digestive system. They also act on the nervous system by invigorating and providing balance. Remedies made using angelica are suitable for people who are overworked and stressed. However, hemlock, an extremely toxic plant with a similar size and the same clusters of white flowers, can be mistaken for angelica. Therefore, if it is gathered in the wild, it is essential to have it checked by an expert to make sure it is angelica.

ANGÉLIQUE

GENRE DES OMBELLIFÈRES

ANGELICA ARCHANGELICA

GREEN ANISE
Pimpinella anisum

Native to the Middle East, green anise has been used since ancient times by the Greeks, Romans, Chinese and Indians. The *Capitulare de Villis*, issued in the early ninth century by Charlemagne, advocated the cultivation of this plant throughout the empire. By the nineteenth century, different species were grown in the French regions of Touraine, Albigeois and Alsace. Green anise production today is mainly concentrated in southern Europe and Turkey. The seeds of this plant, known as aniseeds, together with those of *Carum carvi* (caraway), cumin and fennel, were known as the 'four greater hot seeds'. These plants were believed to heat the body, cure digestive problems, assist sleep, fight bad breath, alleviate palpitations and aches, and encourage lactation in young mothers. Green anise also has other uses, which have given it certain fame. Dragées de Flavigny are a confection made from sugar-coated aniseeds, while a compound found in the essential oil of green anise, anethole, is the base of the famous anisette liqueur, a drink with a high alcohol content invented by Marie Brizard in Bordeaux in the eighteenth century.

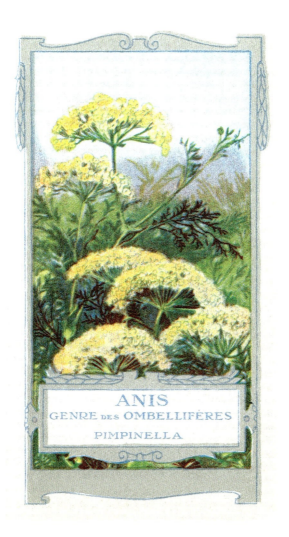

ANIS
GENRE DES OMBELLIFÈRES

PIMPINELLA

STRAWBERRY TREE
Arbutus unedo

This small tree grows wild on the Mediterranean coast and to a lesser extent on the coasts of Brittany and Ireland. It is evergreen, with white bell-shaped flowers and orange-red fruit the size of cherries with bumpy skin. Roman physicians barely showed interest in it. According to them, the fruit was inedible and caused migraines. It had a bad reputation for a long time, and it was even considered to cause intoxication and dizziness. Today, science has shown that every part of the tree have astringent qualities: the leaves, bark and fruit can be used in the form of a decoction to fight infection of the urinary tract. The fruit can be eaten fresh, in relatively small quantities, because it sometimes contains alcohol after fermenting on the tree. Excessive consumption can cause constipation. However, the fruit can be used to make delicious jams – although patience will be required to remove all the bumps from its skin.

ARBOUSIER

ARNICA
Arnica montana

With its large orange-yellow flowers, arnica is distinguished from other flowering plants of the same colour by its sets of two small, opposing leaves growing directly from the stalk. It grows in nutrient-poor meadows with acidic humus, preferring mountain areas. It is also known as mountain tobacco, a reminder of the use country dwellers made of its leaves in the past. Because it does not grow in Greece or Italy, arnica was not known to the physicians of antiquity, and no mention was made of it in medical texts until the middle of the sixteenth century. It is only used externally, because it is toxic to the nervous system if eaten. When used in a compress, it soothes muscular ailments and contusions. It also relieves cuts and bruises, hence one of its French names 'herbe des chutes' or 'fall grass'. Because it is difficult to grow, it is gathered in the wild to supply pharmaceutical laboratories. But this type of harvest has also led to its depletion, particularly aggravated by the use of fertilizers that tend to modify its environment. In order to prevent its extinction, arnica has been protected by a European Union directive since 1992. In North America, a relative of the plant, *Arnica chamissonis*, is grown on a large scale for pharmaceutical use.

ARNICA
GENRE DES SÉNÉCIONIDÉES

PTARMICA

ELECAMPANE
Inula helenium

Elecampane is a sturdy perennial plant that can grow to a height of 2 metres. It grows in ditches, wet meadows and gardens, where it is cultivated for its beautiful yellow flowers that bloom in summer. Native to Asia, this plant was introduced to the Mediterranean countries in ancient times, before spreading throughout Europe, where it became naturalized. In the Middle Ages, it was used by physicians to treat asthma, bronchitis, digestive problems, heart ailments and even the plague. It was later found to have cough-suppressant, antiseptic and antibiotic properties, particularly for the treatment of tuberculosis. Of all the parts of the plant, the root has the most effective action because it contains an essential oil known as elecampane camphor. As a decoction, elecampane is used as a remedy for certain respiratory problems such as cough and bronchial spasms, and also for the digestive tract to stimulate bile production. As an ointment, it alleviates fungal skin infections. A cream is made from the ground root, taken from plants at least two years old, and lard is still used by veterinarians to treat scabies in animals.

AUNÉE
GENRE DES COMPOSÉES
INULA HELENIUM

OAT
Avena sativa

The oat is a cereal native to western Asia. In the Neolithic Period, oat, along with wheat and barley, spread to Europe, where it was cultivated. During ancient times, the Romans believed that the Germanic peoples and the Gauls enjoyed long and healthy lives because they ate the oat. Despite this flattering reputation, it was long used for feeding horses. Humans also ate it in the form of porridge after cooking it with water or milk. But the oat is also a medicinal plant. In the Middle Ages, it was used to treat liver abscesses and gout. In the nineteenth century, its fortifying, vulnerary (wound-healing) and diuretic properties were praised. At the time, a poultice comprising the whole grains and vinegar was prescribed to treat backache and rheumatism. Today, it is known that the oat is a good source of energy. It invigorates and balances the nerves, and it relieves insomnia and stress. For this reason, rolled oats, which are rich in vitamin B, are popular with bodybuilders, young children and convalescents. Oat fibre is good for lowering excess sugar in the blood and cholesterol. As a cosmetic, oat milk, available as a cream or oil, is used as a moisturizer for dried and irritated skin.

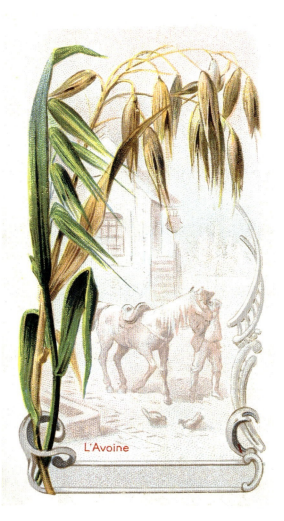

L'Avoine

BURDOCK
Arctium lappa

It is during the summer, when its fruit appears, that the burdock is at its most visible in the countryside. This is the moment when its large capitula or burrs are bristling with small hooked bracts, which cling to the fur of animals and to clothing and hair, to the great amusement of children. These burrs are what inspired the inventor of the hook and loop fastener, Velcro, in 1941. The benefits of this native plant of Siberia have been known for a long time, and they have never varied. Taken as an infusion, the fresh roots and leaves are used to treat ulcers, rheumatism, gout and boils. They also act on all sorts of skin conditions, hence one of the names given to the plant in France, 'herbe aux teigneux' or 'grass of the ringworm sufferers'. In the countryside, burdock was once considered an anti-venom for use after a snakebite or wasp sting. It was used to treat coughing in sheep and scabies on dogs. People also used to eat its long roots, similar to those of salsify, at the end of winter. Burdock deserves to be planted once more in vegetable gardens. But there is one drawback: it is an invasive species.

BARDANE
GENRE DES COMPOSÉES
BARDA

CORNFLOWER
Centaurea cyanus

'The light ... beat down on the large fallow fields covered with cornflowers, turning the chalky soil the palest shade of blue. It clung to the hairs of the flowers and sparkled there ...' The landscape filled with the striking presence of cornflowers, so prettily described by Jean Giono, is nothing more today than a fond memory. These flowers have suffered from the use of chemical products, becoming a scarcity in the countryside. A native of the Middle East, like the poppy, the cornflower spread westwards together with expanding cereal cultivation, hence its name. Any possible medicinal use of the plant seems to have been overlooked in ancient times. It was not until the Renaissance that its antiseptic and anti-inflammatory properties were recognized. It is a good remedy for eye conditions such as conjunctivitis, inflammation of the eyelids and sties. For this reason, it has also been given the name of 'casse-lunettes' or 'spectacle breakers' in France. It is used in the form of a flower water and is applied with compresses. Cornflower water is easy to make. A decoction is made by boiling thirty grams of fresh flowers in one litre of water for five minutes.

CENTAURÉE BLEUET

GREAT MULLEIN
Verbascum thapsus

At the height of summer, this sturdy plant grows mainly in uncultivated areas, unfurling its thick, whitish and downy leaves – known as a substitute for toilet paper in the wild. Row after row of yellow flowers rise up in a single spike, stiff as a candle. In the past, great mullein flowers were added to the bouquets of wild herbs that would be blessed on the feast of the Assumption, 15 August, hence its popular French name 'cierge de Notre-Dame' or 'Our Lady's candle'. The ancient Greeks and Romans already knew of the medicinal properties of this plant. The great mullein is included among the pectoral plants, along with the marsh mallow, common mallow, coltsfoot, violet, red poppy and mountain everlasting. They are all renowned for their expectorant and emollient properties, their ability to soothe a barking cough and alleviate cold symptoms, and for curing ailments of the respiratory and digestive tracts. They also make it easier to sleep. Although the leaves and roots are used, it is mainly the flowers that go into remedies, because they have a high content in mucilage, a viscous substance produced by plants that soothes irritation. They should be picked as soon as they open and quickly dried in a well-ventilated place. A word of caution: when used to make an infusion, they should be very carefully strained because they contain small, highly irritating hairs. As a poultice, the leaves boiled in milk can be used to treat burns and chilblains.

BOUILLON BLANC

GENRE des SCROPHULARINÉES

VERBASCUM THAPSUS

BORAGE
Borago officinalis

As early as April, the borage plant produces small star-like blue flowers that are the delight of bees. Borage grows near homes where the soil is rich in nitrogen. Its thick, dark green leaves are covered in rough and prickly bristles. Probably originating in the Middle East, borage is particularly abundant in the south of France, where it is grown for food: Olivier de Serres, a famous sixteenth-century agronomist, praised its pronounced cucumber-like flavour. Nowadays, borage flowers are a feature of the most sought-after dishes of Michelin-starred restaurants. But there is much more to the plant. The leaves and stalks have diuretic, emollient, sudorific and depurative properties. When prepared as a poultice, they can also be used to treat respiratory ailments, certain kidney conditions and gout. High in fatty acids, borage seed oil is used to lower cholesterol and has an anti-ageing effect.

BOURRACHE

GENRE DES BORRAGINÉES

BORRAGO

BELL HEATHER
Erica cinerea

Bell heather, *Erica cinerea*, and common heather, *Calluna vulgaris*, both belong to the Ericaceae family and have very similar medicinal properties. Both plants have small leathery leaves and pink or mauve flowers. They grow in groups on acidic and mineral-deprived soils, where they form the typical heath vegetation. They have been used since the time of the Renaissance to treat kidney stones. However, it was not until the early twentieth century that these plants were actually found to have diuretic and antiseptic effects on the urinary tract. As a decoction, heather is used to alleviate cystitis, an inflammation of the bladder. The active ingredients are mainly found in the flowering tops of the plant, which are gathered at the start of summer for bell heather and at the end of the season for common heather. Also from the same family is the bearberry, *Arctostaphylos uva-ursi*, which is even more effective in treating urinary tract and kidney ailments, owing to its strong astringency.

BRUYÈRE.

ROMAN CAMOMILE
Chamaemelum nobile

———————————————

The popular name of Roman camomile bears no connection with its origin, because the Romans were unfamiliar with it. The plant grows mainly in the west of France, on the sandy shores of the Atlantic Ocean. It belongs to a family of plants that are very much alike and can be easily mistaken: German camomile, stinking camomile, corn camomile and dyer's camomile. All these varieties of camomile have very similar medicinal properties. The perennial Roman camomile has trailing stems, finely dissected leaves and domed flower heads with yellow disc flowers in the centre and white ray flowers on the outside. Camomile flowers are picked in dry weather at the start of the summer. They have a pleasant smell, although their flavour is a little bitter. Little was known about the camomile plant until the sixteenth century. Soon after its benefits were discovered, it was grown in gardens along with other plant varieties. Infusions made with the dried flowers are known to stimulate appetite, aid digestion and help expel intestinal gases. They are also recommended for relieving headaches and sore throats. Compresses soaked in camomile water are suitable for soothing conjunctivitis and whitlow. As a beauty treatment, camomile shampoo is known for lightening the hair.

CAMOMILLE,
GENRE DES COMPOSÉES

CAMOMILLA

HEMP
Cannabis sativa

Hemp, also known as cannabis, is an annual plant that can grow to a height of 1.5 metres. It has finger-shaped leaves and very small flowers, similar to those of nettle. There are many varieties of hemp. Before it was considered a drug that is banned in most countries, hemp had long been grown, with some varieties used to make rope owing to the qualities of their fibres, while others were used to make clothing or the stiff bedsheets that were handed down from mother to daughter. Protein-rich hemp seed cakes, leftover after processing different several varieties, were used to feed livestock. The flowering tops were also used to extract an essential oil for food use, which is still sold today. However, cannabis varieties that have a high resin content both affect mood and alter the mind. They cause their users a certain form of intoxication, considered to be uplifting, before plunging them into a deep or even hallucinatory sleep. But they also have medicinal properties. They are mainly recommended for alleviating neuralgia and rheumatic pain. Their fruit, hemp achenes, can effectively lower cholesterol if eaten regularly.

Le Chanvre

GREATER CELANDINE
Chelidonium majus

A close relative of the poppy, the greater celandine is distinguished by its yellow flowers. It is typically found in empty plots and flowers with the arrival of the swallows. In fact, it is from this bird, *chelidon* in Greek, that its scientific name and also one of its popular names, swallowwort, are derived. Less poetic is one of its French names 'herbe aux boucs' or 'goat grass', owing to its nauseating odour. The widely extended greater celandine was known to the ancient Greeks and Romans, who attributed many medicinal properties to it, later spread by popular beliefs. It was long recommended for the treatment of eye conditions, the reason for another of its French names, 'grande éclaire' or 'great light'. The juice extracted from its stalks was also considered to have the ability to cure jaundice and other liver ailments. Today, none of these properties have been proven by science. Regardless, greater celandine continues to feature in several of our grandmothers' remedies. Used externally, it is recommended for the effective removal of stubborn warts. For internal use, great care is required because the plant, especially the part underground, contains powerful alkaloids, all of which are toxic. Therefore, greater celandine must only be used once dried, and never in excess of the prescribed dose.

LES PLANTES UTILES

CHÉLIDOINE

COUCH GRASS
Elytrigia repens

———

The stiff stalks, tufts of tall, flat leaves (growing to 1 metre tall) are less characteristic of this grass then its creeping rhizomes, which propagate in all directions and reproduce abundantly. These qualities are the reason for another of its names, quick grass. Couch grass arouses mixed feelings. Farmers and gardeners consider it a weed that is difficult to remove, whereas herbalists prize it for its medicinal properties. Taken as an infusion, couch grass actually has diuretic, depurative, antiseptic, emollient and sudorific properties, which had already been pointed out by the Greek Dioscorides and the Roman Pliny the Elder, two great naturalists of antiquity. The rhizomes contain a great deal of carbohydrate, mucilage, the B vitamins and vitamin A. In the past, country dwellers would resort to couch grass rhizomes in times of famine for making into flour and then bread that offered little nourishment. They could also be used as a coffee substitute or made into sugar and even beer. Dogs instinctively appreciate couch grass, eating it when there is a need to purge themselves, hence another of its names, dog grass.

———

CHIENDENT
GENRE DES GRAMINÉES
TRITICUM REPENS

FOXGLOVE
Digitalis purpurea

The highly characteristic tubular crimson flowers of foxglove, almost all hanging on the same side of the spike, give the appearance of the fingers of gloves. Hence its name, and some of the names by which it is known in France, such as 'gant de Notre-Dame', 'Our Lady's glove', and 'doigt de la Vierge', 'the Virgin's finger'. Foxgloves grow in clumps or drifts, and in spring masses of the plants create gorgeous floral carpets in areas of semi-shade on sandy slopes or at the edge of woodlands. But its beauty is matched by its toxicity. Its leaves contain dangerous molecules, heterosides, which can be fatal. Foxglove is, however, an excellent cardiac stimulant, an agent that increases the strength of the heart's contractions, which only doctors can prescribe in drug form to regulate heart rhythm. William Withering, an English doctor from the late eighteenth century, was the first to discover the plant's benefits for the heart and discovered the digitoxin contained in its leaves. Researchers later learned to control the molecule and turned it into an effective treatment.

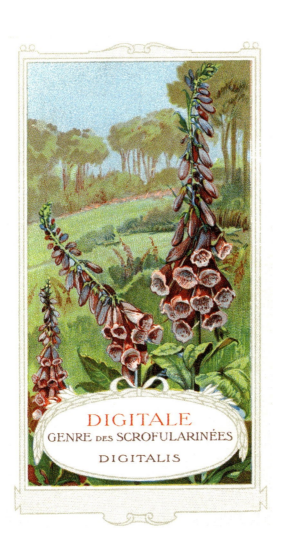

DIGITALE
GENRE DES SCROFULARINÉES

DIGITALIS

COMMON BARBERRY
Berberis vulgaris

This shrub with sharp thorns is often used in gardens to form hedges. It produces clusters of small yellow flowers in late spring, which go on to form round, yellow or red berries, depending on the variety, in autumn. In the past, women would polish its soft wood to use in making items of marquetry. They would also use its bark to prepare infusions as tonics and to aid digestion. They would use the fresh, tart and sweet berries to make jellies, jams and fermented drinks rich in vitamin C. They also pickled them in the same way as capers. The yellow roots and red fruit were also used to make dyes for cloth. Farmers, on the other hand, were very suspicious of the shrub because they believed it would spread rust, a fungal disease that could quickly wipe out an entire crop. Science would later prove them right. Today, we know that, except for the berries, the entire shrub is toxic. It contains berberine, a powerful alkaloid. However, when under strict medical control, an infusion or decoction of the bark alleviates digestive problems and acts as a stimulant in case of severe fatigue.

ÉPINE - VINETTE

GENRE DES BERBÉRIDÉES

BERBERIS VULGARIS

TASMANIAN BLUE GUM
Eucalyptus globulus

In 1792, Bruny d'Entrecasteaux's mission to the Pacific in search of Lapérouse and his missing expedition led him to Tasmania and the discovery of the Tasmanian blue gum. This eucalyptus species grows very tall and can reach a height of 50 metres. Its trunk sheds bark, and its leaves are thick and leathery. The youngest leaves are a bluish colour and the older leaves are a deep green. Today, this tree has adapted to life in many parts of the world, particularly the Mediterranean region. It is exploited for its wood and essential oils, however, it has caused some environmental problems. Its leaves do not decompose, so they form a dry litter with little organic value that turns the soil sterile. Eucalyptus trees are also extremely prone to fire. Their main active ingredient, eucalyptol, is found in the leaves harvested from the older branches and dried. Used in the form of a suppository, lozenge, infusion or syrup, eucalyptus is an effective treatment for asthma and bronchitis, and it soothes coughs. Activated charcoal made from eucalyptus wood is recognized as a treatment for intoxication because of its ability to absorb all forms of toxins from the intestines.

EUCALYPTUS

GENRE DES MYRTACÉES

EUCALYPTUS GLOBULUS

FENNEL
Foeniculum officinalis

Native to the Mediterranean region, fennel spread towards India and China. In France, it was grown in monastery gardens in the Middle Ages. Today, it is known as an aromatic and culinary plant and is used to add flavour to fish, for instance. It also found in the wild, where it grows spontaneously. On hot days, its strong smell of anise can be perceived on dry slopes exposed to the sun. Fennel is identified by its long striated stalks, finely dissected leaves and large umbels of tiny yellow flowers. It was once believed to be able to cure all sorts of physical and mental ailments, including melancholy. It was included in the list of 'four greater hot seeds', plants considered to stimulate the appetite and to aid flatulence. It was also included among the ingredients of the 'syrup of five roots' and age-old diuretic preparation, along with smallage, asparagus, parsley and butcher's broom. Today, science has acknowledged it mainly for its digestive, expectorant and skin-healing properties. Fennel honey is recommended for treating rheumatism and, for chronic conjunctivitis, an infusion of fennel seeds has been proven effective for cleaning the eyes.

FENOUIL

GENRE DES OMBELLIFÉRES

FOENICULUM VULGARE

JUNIPER
Juniperus communis

The juniper is a small, thorny and bushy conifer that grows on lime-rich hillsides, acidic heaths and in peat bogs – its only requirement is sunlight. In the 1960s, it was still found in abundance in open pasture land left for the grazing of goats and sheep, which naturally avoided its thorny branches. Nowadays, as pastures are abandoned, juniper is becoming scarce, endangering a plant that humans have been using since the earliest times. In the Neolithic Period, the aromatic ashes of its branches were already being used to smoke meat. Much later on, country dwellers would instead seek juniper for its fleshy seed cones, known as berries, which were used as a seasoning for sauerkraut or as ingredients for digestive drinks, ginever and gin. Since then, science has shown that these berries, when eaten, have diuretic, invigorating and expectorant properties. They also stimulate the appetite and aid digestion. A word of warning: pregnant women and people who suffer from infection or inflammation of the kidneys are strictly advised against taking infusions and decoctions of juniper. Rubbing with juniper essential oil is recommended for providing relief from osteoarthritis and rheumatic pain.

GENÉVRIER

GREAT YELLOW GENTIAN
Gentiana lutea

The great yellow gentian grows in the alpine regions of Europe and the Middle East, where it produces yellow flowers along its sturdy stalk. It is accustomed to harsh winters, short springs and sunny summers. It has been known since ancient times. The Greeks and Romans attributed to it countless qualities, including being an antidote to snakebite, a remedy against the plague, a treatment for liver ailments, among others. Before the discovery of quinine, in the mid-seventeenth century, it was used to reduce fever. Today, the great yellow gentian is disappearing. Because its root is so highly sought after for its medicinal properties, the plant has suffered from overharvesting. When taken internally, gentian is known to stimulate circulation of the blood and a lazy gall bladder. It is also known for stimulating the appetite and aiding digestion, which is why it is used in a many liqueurs and alcoholic drinks made in mountainous regions (such as Auvergne, Jura and Savoy).

LES PLANTES UTILES

GENTIANE

HERB ROBERT
Geranium robertianum

Herb Robert is a member of the geranium family that is commonly found along shady paths and on the edges of woodland. Its denomination is not derived from the name Robert, but from the Latin *ruber*, 'red', the main colour of its stalks. When rubbed between the fingers, its leaves give off a pungent smell produced by its essential oil. It has pinkish-red flowers and its fruit resembles the bill of a crane, the origin of one of its French names, 'bec de grue'. In the eighth century, the bishop of Salzburg discovered this geranium's haemostatic properties, its ability to stop bleeding. It was also attributed with the ability to cure kidney stones, inflammations of the breast and mouth ulcers. However, today, only the external uses of the plant are recognized. As a lotion, herb Robert is used to treat conjunctivitis; as a mouthwash, it is effective against gingivitis and stomatitis; and as a poultice, it aids the healing of skin infections.

HERBE A ROBERT
Geranium Robertianum

GINGER
Zingiber officinale

Ginger is a perennial plant with blade-like leaves that is native to Asia, either China or India. It thrives in hot and humid climates. Its name is derived from the Sanskrit, *shringavera*, meaning 'deer antler', in reference to the rhizome or root. In the first century AD, the Greek physician and botanist Dioscorides advocated its use for the treatment of stomach ailments. In the Middle Ages, Marco Polo made an accurate description of the plant during his travels in China. Arabs and Persians were then importing ginger by way of the caravans that travelled the Silk Route or by the boats that traded along the Arabian coast. At the same time, soldiers on the Crusades also showed interest in it, bringing it back to Europe, where it would be used as a seasoning to replace the extremely expensive spice pepper. Ginger is still popular for flavouring in Great Britain, the Unites States and Australia, where it is used in pastries and cakes such as gingerbread. It is also found in other cuisines around the world, as evidenced by the Indian garam masala, Japanese sushi and Canadian *ginger ale*, among others. In the sixteenth century, ginger continued on its journey to the New World, where it was cultivated. Whatever the time or country, ginger is particularly sought after for its root, with a spicy and fiery flavour. It is used as a carminative, to relieve flatulence; it calms nausea and vomiting; and it aids digestion.

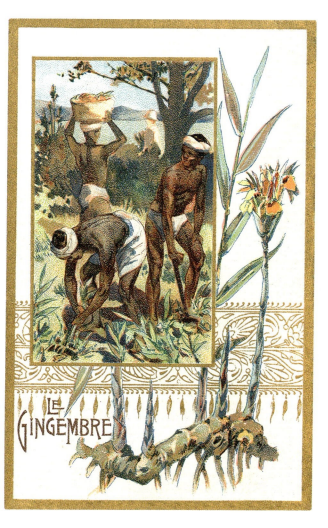

LE
GINGEMBRE

MARSH MALLOW
Althaea officinalis

The marsh mallow is a large perennial plant with pretty, pale pink, five-petal flowers and velvety, downy leaves. It grows in the rich soils of marshes and wet meadows, mainly close to the coast. It is believed to have originated on the steppes of Central Asia, from where it gradually spread westwards. In the first century AD, the Greek physician Dioscorides attributed all sorts of benefits to it, some unexpected, such as aiding childbirth. In the Middle Ages, the marsh mallow was found in the ordinary gardens of monasteries and in country dwellers' gardens, where it was grown as a remedy for migraines. Today, it is known that the entire plant, particularly the root, contains minerals and vitamin C. It is especially high in mucilage, which makes it suitable for ailments of the lungs and a good emollient. Because of these natural softening agents, a decoction sprayed onto the skin makes it softer. Taken as an infusion, it is renowned for soothing throat irritations, dry coughs and different types of inflammation. In the past, teething babies were given marsh mallow root to chew on. This remedy has also been effective for soothing inflamed gums.

LES PLANTES UTILES

GUIMAUVE

COMMON HOP
Humulus lupulus

The hop is a climbing plant typically found growing in hedges bordering streams and in cool and damp woods. It produces female flowers in the form of rounded, hazelnut-like cones that hang in clusters. The plant grows to a length of 5 or 6 metres, and its stalk winds around a supporting shrub. The hop is cultivated in regions with a temperate, mild and humid climate. Since around the eleventh century, its inflorescences, known as hops, have been used in the production of beer, providing its well-known bitter flavour. The hop has long been attributed many medicinal properties, some of which are being questioned today. In the twelfth century, it was thought to be effective for the treatment of melancholy, which is highly unlikely. However, the hop is used for cases of painful menstruation and for all sorts of disorders affecting women after the menopause. It is also thought to have an oestrogenic effect. As far as men are concerned, its properties for reducing sexual appetite have been proven, particularly with excessive consumption of beer. It is known that the hop strengthens the stomach and soothes anxiety. It is also recommended for insomnia. There are two practical ways it can be used to help you sleep: by placing a pillow filled with hops under your head, or by drinking an infusion of hops before going to bed.

HOUBLON

BAY TREE
Laurus nobilis

Most likely originating in Asia Minor, this evergreen tree has leathery leaves and fruit resembling small olives. It is also known by a long list of other names, including true laurel, victor's laurel and poet's laurel. Each of them is a testament to the use and importance given to the bay tree throughout history, from its association with the Greek god Apollo to its use today as a herb. In ancient times, a laurel wreath made from its leaves was placed on the head of victors as an honorary distinction and symbol of glory. It became a mark of Roman emperors to represent their authority over the army. The bay tree was also attributed with the power to guard against lightning, to purify the air and to ward off evil spirits. It was considered a panacea in the Middle Ages, with the power to cure all sorts of ailments, even the most unlikely ones, such as obstruction of the spleen and angina. Nowadays, despite the bay tree having been relegated to the rank of simple culinary plant, guaranteeing the success of a bouquet garni, its true medicinal properties should not be overlooked. When made into an infusion, it stimulates the appetite, acts as an expectorant for bronchitis and aids digestion; in the form of a cream, oil or balm, it alleviates rheumatism and stiff neck.

LE
LAURIER SAUCE

ENGLISH LAVENDER
Lavandula officinalis

Everybody recognizes lavender and its exquisite, penetrating fragrance, its flowers reminiscent of the sunny and rocky hills of Provence. Although it grows wild on the dry, lime-rich soils on hillsides at medium altitude (400–1,800 metres) in the western Mediterranean, today it is mainly cultivated in compact rows. But that commercial variety is in fact lavandin or French lavender, a sterile hybrid of English or true lavender (*Lavandula vera*) and spike lavender (*Lavandula spica*). In ancient times, the Romans loved to scent their bathwater with lavender. This use is what gives the plant its name: lavender comes from the Latin word *lavare*, which means 'to wash'. The uses for this plant are endless. In small sachets, it repels lice, mites and parasites, and adds its fragrance to household linen stored in cupboards. Taken as an infusion, it stimulates the appetite and aids digestion. In this form, it also serves as a sedative and soothes migraines. A sachet of lavender placed under the pillow and massaging the temples with lavender essential oil have the same effects. Used externally, in the form of a tincture dissolved in alcohol or oil, lavender has skin-healing properties; it serves as a balm for bronchitis sufferers, and it offers relief for gout, insect bites, burns and rheumatic pain. Another of its properties is that it lifts the spirits and alleviates stress.

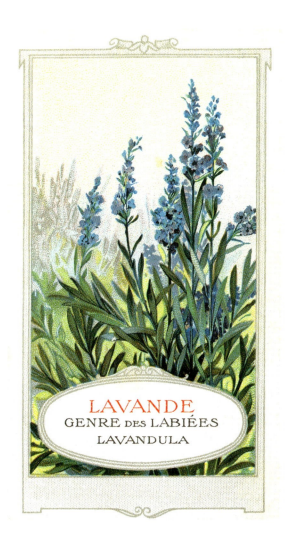

LAVANDE

GENRE DES LABIÉES

LAVANDULA

GROUND IVY
Glechoma hederacea

The only ivy in this plant is its name! Except for the way it creeps over the ground, it bears no resemblance to its namesake, which tends to climb over walls. This small and discreet perennial shade plant is commonly found in neglected gardens and abandoned hedges and orchards. It produces small, pretty blue flowers at the height of spring, before the leaves open on the trees. It has creeping stalks, small, round, crenate (tooth-edged) and sometimes reddish leaves, and a penetrating smell, which can be unpleasant. During the Middle Ages, it was an ingredient in folk remedies. It was even considered to have magical powers. People thought it worked on their digestive tract, cured them of jaundice, rid them of intestinal worms and healed nasty wounds. It was even claimed to cure madness! Nowadays, it is mainly known for its properties for treating the respiratory system. When taken as an infusion, it treats chest coughs, bronchial catarrh, asthma and lung diseases with heavy production of mucus. As a compress, it also relieves haemorrhoids.

LIERRE TERRESTRE

GENRE DES LABIÉES

GLECHOMA HEDERACEA

FLAX
Linum usitatissimum

Flax once had a place in all farmers' gardens, together with hemp and a few fruit trees. The plant was considered precious, and for good reason. It produced a fibre that was woven into linen, used for garments, household cloths and sheets. Artists prized it for the solid and firm canvas that it made for their oil paints, giving incomparable brilliance to their creations. The flax plant also produces seeds, linseeds, with medicinal properties. The ground seeds sprinkled with mustard powder are used in hot poultices that are applied to the body to relieve different inflammations of the respiratory system, digestive tract and joints. Moreover, these seeds make an excellent remedy for constipation. Today, fresh linseed oil is often used as a laxative.

LES PLANTES UTILES

LE
LIN

MAIZE
Zea mays

Maize, also known as corn, is native to the Americas. It was brought back to Europe by the Spanish in 1520. It gradually spread throughout the continent, where it was given different names, such as Spanish corn, Indian corn and Turkish corn ('corn' at the time meaning grain, before becoming a synonym for this plant). Later, when it had become naturalized in the south-west of France, it was given the name 'French corn'. It was soon considered providential food: made into cakes and porridge, it provided nourishment for families, helping them to overcome times of famine. Unfortunately, because maize was the only thing they ate, country people became prone to pellagra, from the Italian for 'sour skin', which appeared as afflictions of the skin and types of dementia. Today, maize has been proven to be a beneficial plant. Gluten-free maize flour makes a good substitute for wheat flour. It protects the intestinal mucosa and helps to restore the health of anaemia sufferers. Corn oil is high in unsaturated fatty acids and lowers cholesterol. Corn silk, the fine hair-like tassels found on the ears of maize, also has curative properties. It is highly diuretic and cleansing, making it very useful for relieving swollen legs, gout and arthritis.

LES PLANTES UTILES

LE MAÏS

WHITE HOREHOUND
Marrubium vulgare

Commonly found in the south of France, where it grows in rubble, on highly exposed hillsides and on the edges of gardens, white horehound is a little plant with many good qualities, which has certainly earned it one of its common names in French 'bonhomme' or 'good fellow'. In ancient times, it was used to treat a number of respiratory ailments. The Greek physician Dioscorides said at the time that it 'brings up congestive matter from the chest'. The monks of the Middle Ages who practised herbal medicine and the physicians of the following centuries saw in it these same properties, and a number of others for the treatment of ailments of the liver, jaundice, worms and swelling of the breast, among others. Today, the qualities of white horehound have been better identified and, for the most part, confirmed by science. The plant makes a particularly good expectorant, calming coughs and loosening and eliminating mucous secretions. It can also be used to stimulate the appetite, and it is recommended for people who have an aversion to food. The freshly picked flowering tops make a good herbal tea, sweetened with honey.

MARRUBIUM VULGARE.   MARRUBE COMMUN.

COMMON MALLOW
Malva sylvestris

The common mallow can be found everywhere in summer, along hedgerows and paths, on fallow fields and in woodland clearings. Its pretty mauve flowers with dark purple veins are continuously turned towards the sun. The ancient Greeks and Romans were bewitched by the flower, and they sowed mallow seeds near tombs in order to provide the dead with serenity and eternal peace. The plant was eaten as a vegetable. The well-known Roman author Cicero, who loved to feast on mallow stews, unfortunately, had to experience their unpleasant laxative effects afterwards. At that time, it was also taken as an infusion or a decoction, or used in a poultice to treat different ailments, mainly inflammations. The admiration it was shown in ancient times did not lessen with the years. Olivier de Serres, a famous sixteenth-century agronomist, considered it to be 'singular for the treatment of the teeth'. It was also recommended for easing constipation in children and adults. It was believed that a single morning tisane inspired its drinker with serenity for the entire day and protected against illness. Today, it is known for certain that the common mallow contains a good deal of mucilage, a natural softening agent. In fact, it has a beneficial effect on coughs. Taken internally, it is a sedative and laxative. Used externally, as a mouthwash, decoction or poultice, it works wonderfully as a treatment for ailments of the skin and mouth.

LES PLANTES UTILES

MAUVE.

LEMON BALM
Melissa officinalis

Native to Asia, this plant today is commonly found in European gardens. Its wrinkled leaves give off a fine and very pleasant fragrance of lemon and geranium, and it provides bees with its scented nectar. Although it was known to Dioscorides and Pliny the Elder, it was mainly the Arabs, from the tenth century onwards, who recommended it as a tonic able to ward off insomnia and worries, and to bring joy and calm palpitations of anxiety. In the eleventh century, Avicenna, the famous Persian physician, once asserted that lemon balm 'makes the heart merry and joyful'. In the seventeenth century, Parisian monks used it as a flavouring for their famous Carmelite water. This herbal tonic, which acted on the heart, was the result of distilling a mixture of fresh lemon balm, alcohol and white wine, flavoured with a secret number and quantity of spices. For a long time, it was popular with ladies, who made good use of it as a digestive. Today, for cases of slow digestion and bloating, an infusion of dried lemon balm leaves is recommended. This plant has also proven to be effective for problems with sleep, stress, depression and anxiety. Used in a compress, it is recommended for the treatment of herpes.

MÉLISSE

GENRE DES LABIÉES

MELISSA OFFICINALIS

MINT
Mentha pulegium, M. piperita and *M. viridis*

Everybody is familiar with mint and its characteristic scent. There are countless species of this plant that are difficult to tell apart. Among them, pennyroyal, *Mentha pulegium*, has long been known to have medicinal properties. In the fifth century BC, the famous Greek physician Hippocrates considered it to be an aphrodisiac. His successors would mention it as a remedy for snakebite, weariness, cough, cholic, nausea and fever, among others. Moreover, it was enough to burn a few fresh flowers in order to clean places infested by fleas. This insecticidal property earned pennyroyal its scientific name *pulegium*, the Latin word for 'flea', and its popular name in French 'chasse-puce' or 'flea deterrent'. Today, pennyroyal is mainly used as a digestive tonic. Its use is recommended with steam inhalation to treat persistent cold or to alleviate ailments of the mouth. Peppermint (*M. piperita*) and spearmint (*M. viridis*) have practically the same properties. Fresh mint leaves also flavour Moroccan tea and repel mice, which cannot stand the smell. They also prevent the curdling of milk through a chemical reaction.

LES PLANTES UTILES

MENTHE

ST JOHN'S WORT
Hypericum perforatum

Growing along the side of paths, St John's wort has been known as the plant of a thousand holes, owing to the scattering of tiny translucent glands over its leaves. Its name comes from the tradition that it should be picked at noon, the warmest time of the day, on 24th June, the feast of St John the Baptist. Since ancient times, St John's wort has been attributed to have wonderful powers that strike the imagination, including warding off demons and restoring order to the home. Today, science has shown that the flowering tops, with bright yellow blooms, have astringent and antiseptic properties, which are useful for treating and healing the skin for burns, sunburn and insect bites. St John's wort is effective when the fresh flowers are macerated in an oil, which turns red, then applied in this form. However, the skin must be kept out of direct sunlight because the plant contains a substance, hypericin, that is sensitive to light and may cause inflammation. The plant is used in Germany to make an infusion to treat anxiety, nervousness and mild depression. It is also known as a sleep aid.

BITTERSWEET
Solanum dulcamara

This thin and supple vine grows in semi-shade near water, in hedgerows and on lightly wooded land, preferring rich soils. It is also common in gardens. It produces star-shaped purple flowers with a yellow centre. In autumn, it bears red berries similar to redcurrants. Although birds adore them, they are toxic to humans, and if eaten can cause diarrhoea and serious problems to the nervous system. When chewed, the leaves and young stalks have a flavour that is first sweet and then bitter, giving the plant its name of bittersweet. In the past, all parts of the plant were used. The thin stalks were woven into baskets; its fruit was used to make green and purple dyes; its branches were used to treat bronchitis and reduce fever, which is the reason for one of its common names in France, 'herbe à la fièvre' or 'fever grass'. Its soft stems were used to cure skin conditions, insect bites and rheumatism, among others. It was also believed that by boiling the leaves with bacon and applying them as a poultice, they could quickly heal the nastiest of tumours. Today, given its toxicity, bittersweet is only used externally under strict medical control to heal wounds and ulcers in the form of decoctions and poultices.

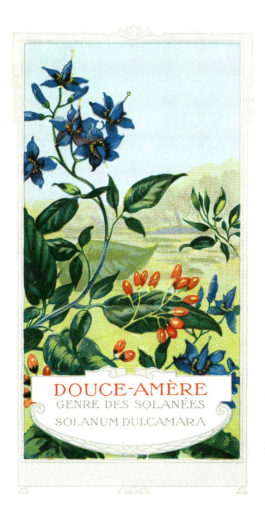

DOUCE-AMÈRE

GENRE DES SOLANÉES

SOLANUM DULCAMARA

BLACK MUSTARD
Brassica nigra

Black mustard, *Brassica nigra*, is a honey-producing annual plant. It is commonly found and mainly grows near water. Its flower comprises four yellow petals, and its leaves have a spicy flavour. Its tiny, black and fiery seeds are used to make the famous mustard plaster. This sort of poultice, made of mustard powder mixed with linseed flour, is applied hot to the chest or back to treat a bad case of bronchitis, to alleviate rheumatic pain and inflamed joints, and to provide relief from sciatica, among others. It should be noted that prolonged use can cause the skin to blister. The seeds are also digestive and stimulate the appetite. The large white seeds of the white mustard plant, *Sinapis alba*, which are used to make the famous condiment, have similar, although less intense, properties.

LA
MOUTARDE

MYRTLE
Myrtus communis

The myrtle is a delightful plant, with its fragrant white flowers that scent the exposed hillsides around the Mediterranean between May and July. Its fruit – small, dark and dry berries – have a very pleasant resinous flavour. Its leathery, evergreen leaves are dotted with small translucid glands with the pretty French name of 'eau d'ange' or 'angel water', which are rich in essential oil. Since the dawn of time, humans have turned to this plant in their search for well-being. It was mentioned in the Old Testament and was sacred to the Persians. It was a symbol of beauty and youth for the ancient Greeks and Romans, who offered branches of myrtle to the goddess Venus. Greek, Roman and, later, Arab physicians considered the myrtle a panacea. Its leaves and berries, which were made into a fortified wine, were used to darken the hair of women, soothe coughs, mend badly broken bones and season poultry dishes, among other uses. Today, the myrtle is recognized to have antiseptic and astringent properties. Taken as an infusion or used as a mouthwash, myrtle relieves respiratory and urinary conditions, and is used to treat abscesses and boils.

Le Myrte

OLIVE TREE
Olea europaea

The quintessential source of sustenance, this tree is redolent of the Mediterranean, with its warm, rocky landscapes and light-filled evenings, and its symbolic value is as strong as that of wheat and the almond tree. In Christianity, it represents peace, mercy and the Holy Spirit. It is found everywhere, from Egypt and Asia Minor, its original homeland, throughout the Mediterranean region, and as far as the Americas, where it was introduced by the first settlers. The olive tree is naturally renowned for its fruit, olives, which are rich in vitamins, stimulate the appetite and aid digestion. For a long time, olives accompanied by a piece of hard cheese made a typical shepherd's meal. Olives are also a part of the famous Mediterranean diet, which reportedly ensures health and long life, and prevents the risk of heart attack. Virgin olive oil, rich and green, and preferably cold-pressed, is an emollient. It nourishes, scents and softens the skin, and heals burns and sunburn. It also increases the levels of good cholesterol in the blood. The leaves, with febrifuge and hypotensive properties, are recommended in infusion for reducing fever and blood pressure.

LES PLANTES UTILES

L'OLIVIER

ORANGE TREE
Citrus vulgaris

A native of southern China, the orange tree, *Citrus vulgaris*, spread to India before reaching Persia, where it was known as the Median apple. The ancient Greeks and Romans came to know of it quite late. The bitter or Seville orange was the first variety to be grown in Europe, introduced by the Arabs to Andalusia and Sicily in the eleventh century. Four centuries later, it was mainly the Portuguese who cultivated and developed the sweet orange. In the age of great discoveries, the tree conquered the New World, where it grows in Mexico, Florida and California. The bitter orange is rich in active ingredients. Taken as an infusion, its leaves and flowers have sedative properties, helping to bring sleep, calming nervousness and migraines, digestive problems and heart palpitations. The peel has the properties of a tonic, digestive aid and appetite stimulant. It is used in the production of liqueurs, such as curaçao, Grand Marnier and Cointreau. The distilled flowers are the basic ingredient of orange flower water, which gives relief from pain, and also of eau de Cologne and Hungary water. The sweet orange is an excellent source of vitamin C, mainly turned into juice. In the past, it was used to treat scurvy. Because it is a tonic, it is recommended for people suffering from fatigue and anaemia.

Le Monde des Plantes

Famille des Rutacées

1. ORANGER 2. RÜE FÉTIDE

OREGANO
Origanum vulgare

Oregano grows on warm and sunny slopes. Its scent heralds the arrival of August. You need only rub a sprig between your fingers to release its aromatic and balsamic flavours. In prehistoric times, hunter-gatherers would use it to season fresh meat. Today, oregano is more commonly used to flavour grilled meats and pizzas. In the countryside, it was commonly used for its aromatic and therapeutic properties, some of which are still recognized. The plant served as a condiment for cooking, as a tonic to calm a persistent cough, as a remedy for aerophagia and leprosy, and as a lucky charm to ward off demons and witches. Compresses made from the freshly picked flowering tops, which were heated and directly applied, could even cure a stiff neck. This list of benefits explains why each summer people would make small bundles of oregano, which were then left to dry in the attic. Today, the flowering tops of oregano are more commonly taken as a light tisane, known as 'red tea' by its colour. This drink stimulates the appetite; it eases difficult digestion, bronchitis, toothache and headache; and it calms a dry cough. A relative of oregano, marjoram, *Origanum majorana*, is mainly grown in gardens for culinary use.

ORIGAN
GENRE DES LABIÉES
ORIGANUM VULGARE

WHITE DEADNETTLE
Lamium album

Lamium album is known as the white deadnettle, because unlike the common or stinging nettle, from the *Urtica* genus, it does not sting. It is more closely related to the herbs mint, garden thyme and wild thyme, and rosemary, and gives off its scent when rubbed between the fingers. It has small, white, lip-shaped flowers, which are easily seen and allow it to be easily identified. It grows near homes, on rubble heaps, in close proximity to livestock and stables, but also in damp woodlands, in the company of ramsons and bluebells. While it was given little use prior to the fifteenth century it later took pride of place in folk medicine. Its flowers, at the same time tonic and astringent, diuretic and refreshing, would be used to treat coughs and diarrhoea, swelling and spitting blood, varicose veins and bruising. Today, all those properties have been confirmed by medicine, in particular the plant's anti-inflammatory action on the skin and mucosa when used as a compress, and its ability to treat urinary and digestive tract ailments and sleep problems when taken as an infusion. The flowering tops in a bath will relax a rambler's tired feet.

ORTIE BLANCHE

PASSION FLOWER
Passiflora incarnata

In the sixteenth century, Jesuit priests arriving in Peru in the wake of the Spanish conquistadors, named this long vine the passion flower. They saw in the corona of filaments in the flower's centre a representation of Christ's crown of thorns. But their comparison did not stop there. In their view, the ends of the pistils resembled the nails used to hang Christ on the Cross, the stamens, the sponge soaked in vinegar lifted up to Jesus to quench his thirst, and the fruit, the heart of God. The passion flower was brought back to Europe and quickly spread throughout Mediterranean gardens. The aerial parts of the plant, the leaves and flowers, mainly have sedative properties. Taken as an infusion, they ensure quality sleep, inhibit sleeplessness and palpitations, and combat stress and anxiety. The passion flower is also a gentle tranquillizer free from side effects. Its cousin, the passion fruit, *Passiflora edulis*, is a vine that produces invigorating fruit of the same name that are high in vitamin A.

PASSIFLORE

OPIUM POPPY

Papaver somniferum var. *album*
and *P. somniferum* var. *nigrum*

———————————

The opium poppy, *Papaver somniferum* var. *album*, is also known as the dream plant. It produces large flowers with petals ranging in tone from pink to purple. Its fruit, in the form of a large green capsule, is filled with a milky juice, latex, which contains opium. Depending on its use, this natural substance can be seen as a powerful illicit drug or, conversely, as an invaluable remedy against pain. The opium poppy has been grown since ancient times. In the first century BC, the Roman poet Virgil described the flowers as *soporiferum*, meaning 'sleep-inducing', which is evidence of the use it was given at the time owing to its sleep-inducing properties. Barely a century later, Dioscorides, who recommended it to relieve pain, warned of its harmful effects. In the mid-nineteenth century, this poppy was at the heart of the famous Opium Wars between Britain and China. Today, opium and its derivative products must only be consumed under medical control. The black or blue-seeded *P. somniferum* var. *nigrum*, known as the breadseed poppy, is grown for its seeds, which are used to decorate loaves of bread and pastries. Poppy-seed oil is also famed for being as delicate as walnut and almond oil.

———————————

PAVOT
GENRE DES PAPAVÉRACÉES

PAPAVER SOMNIFERUM

WILD PANSY
Viola tricolor

The wild pansy belongs to the same family as the early dog violet. It grows in the fields and meadows, along paths and on thin sandy soils, and flowers for much of the summer. With its four light upper petals and one dark yellow bottom petal marked with purple, it is a little more drab than its cousin. It is also less fragrant. Perhaps, as the legend goes, this is because it begged the Trinity, its patron saint, to weaken its remarkable scent in order to stop country dwellers from picking it. In the language of flowers, wild pansy also symbolizes memory. No ancient or medieval writings attribute any medicinal properties to the wild pansy. It was only in the sixteenth century that physicians began to show an interest in the plant. Today, the flowers and leaves are recommended for their depurative properties, and can be used in the form of a compress soaked in an infusion of the plant to clean the skin. It is mainly used to treat cradle cap and skin conditions such as psoriasis, acne, eczema and dandruff. In cosmetic creams and ointments, wild pansy is reputed to relieve dry skin and stretch marks.

PENSÉE SAUVAGE
GENRE des VIOLARIÉES
VIOLA TRICOLOR

LESSER PERIWINKLE
Vinca minor

The lesser periwinkle has magnificent blue flowers in spring. Its leathery leaves are always green and its long stalks trail along the ground. The striking botanical features of the lesser periwinkle have earned it a number of different common names, such as the French 'violette de serpent' and 'petit sorcier', meaning 'snake's violet' and 'little wizard', respectively. It is said that wizards once made great use of it as an ingredient in their magic potions. It was also used to treat diarrhoea and toothache. In the seventeenth century, its leaves were used as a remedy against lung disease, fever, spitting blood and bleeding. Today, archaeobotanists have established a link between the abundant presence of lesser periwinkles, which love nitrogen-rich soils, and intense human activity (farming, coal mining, etc.). This plant is an ingredient in medicines used to treat memory loss and ageing. It is typically taken as a decoction, which stimulates blood flow to the brain and enhances its oxygenation. It is also used to suppress lactation in women who want to stop breastfeeding.

LES PLANTES UTILES

PERVENCHE

COMMON CENTAURY
Centaurium erythraea

Common centaury grows in dry and sandy places and in meadows, preferably in lime-rich soils. It has the surprising peculiarity of only opening its flowers when the temperature suits it, when it is at least 24°C. The name of the species bears does not refer to any connection with Eritrea, but comes from the Greek word *erythraea*, 'pink', the colour of the petals of its flower. The Greeks, Romans and Gauls once considered the common centaury as a panacea. This reputation of miracle cure continued into the Middle Ages and the Renaissance. Like several plants in the gentian family, it actually has important medicinal properties. It was once used to treat high temperatures, which earned it the name of feverfoullie. As an infusion, common centaury was also appreciated for its invigorating and stimulating effects, and vermifuge (treatment for worms) and antiseptic properties. As a decoction, it was used to heal the skin, repel lice and stop hair loss. Today, all those properties have been recognized by medicine, with one reservation: it irritates the digestive tract.

PETITE CENTAURÉE
GENRE DES GENTIANÉES
ERYTHRŒA

BRANCHING LARKSPUR
Consolida regalis

A native of Asia Minor, the branching larkspur, *Consolida regalis*, is thought to have arrived in Europe during the Neolithic Period, along with cereals. Today, it is a flower that graces our gardens. From one year to the next, it produces its lovely, well-stocked clusters of flowers, which can be blue, white or mauve, depending on the horticultural variety. In the wild, its flowers are deep blue, standing out against the golden colour of the wheat in which it would sometimes grow, until it was considered a weed that needed to be eliminated. The branching larkspur is a perfect example of a medicinal plant from the past rejected by present-day medicine because of its toxicity. Its seeds were once used to make creams to treat scabies, ringworms and lice. Nowadays, it is known that this species of the buttercup *(Ranunculaceae)* family has a high alkaloid content, making it dangerous.

LE PIED D'ALOUETTE

PEPPERS
Capsicum annuum and *C. frutescens*

There are at least three hundred species of peppers. Among them is the chilli pepper, *Capsicum frutescens*, and the sweet red pepper, *C. annuum*, known in France as 'corail des jardins' or 'garden coral', because it turns the colour of a sea creature as it ripens. Peppers were discovered by Christopher Columbus's physician on the island of Hispaniola (today divided between Haiti and the Dominican Republic), who had noticed the natives cooking them as food. They were quickly sent back to Europe, where they were used to season broths and meats. They were used as a remedy against the build-up of pituitous humours, sputum and phlegm, and to relieve sciatica. Today, they are known to be high in beta-carotene and vitamin C, and for their stimulating properties in the case of slow digestion. Used in a poultice, only under medical control, the chilli pepper soothes rheumatic pains, stiff neck and lower back pain. Paprika, a red powder with a flavour that is much sweeter than the spicy and strong flavour of the tiny Indian and American chillies, is made from sweet peppers grown in Hungary.

LE
PIMENT

BURNET
Sanguisorba officinalis and *S. minor*

Although a close relative of the rose, burnet bears no resemblance to it. It produces no thorns or fragrant petals. On the contrary, it has a smooth stalk and small flowers growing in clusters on round heads, which are green in the shade and red if exposed to the sun. There are two species of burnet. The tall and quite rare great burnet, *Sanguisorba officinalis* grows in damp places and sometimes in mountainous regions. The more common salad burnet, *S. minor*, prefers dry grassy meadows and lime-rich soils. They both share the same medicinal properties. They have proven haemostatic qualities, which are particularly well explained by their scientific name, *Sanguisorba*, comprising the Latin words *sorbere*, 'to absorb', and *sanguis*, 'blood'. Because they are rich in different tannins, they are astringent and skin-healing. When used in a decoction, they are very good for treating wounds and ulcers. Taken internally, they are also digestive and diuretic, and are recommended as remedies for diarrhoea and dysentery. Salad burnet is high in vitamin C, and, as its name suggests, can be used in salad in the same way as escarole and Batavia lettuce.

LA PIMPRENELLE

DANDELION
Taraxacum officinale

It is difficult to find one's way around the world of the dandelion. Even seasoned botanists are barely able to distinguish between the species and subspecies because there are so many of them. Regardless, everybody easily recognizes its yellow flowers, the colour of freshly churned butter, its silky clock, which are blown for the pleasure of watching the seeds fly, and its hollow stem, which releases a milky substance. One of its folk names, 'pissabed', is a wonderful description of its diuretic properties, known since ancient times. The dandelion is high in vitamins, iron and potassium, giving it the properties of a tonic and a cholagogue, meaning that it promotes the release of bile. It also decongests the liver. After a winter diet that is high in fat, a treatment consisting of fresh dandelion salad is strongly recommended, because the plant is both a stimulant for the appetite and a depurative. All that is required is a little olive oil and a drizzling of lemon juice. The young flower buds can be used like capers. As for the roots, once roasted, they can become a delicious and, of course, caffeine-free coffee substitute. The flowers, which are used to make jellies and to flavour wine, are also very much at home today in the refined dishes created by Michelin-starred restaurants.

PLANTAIN

Plantago major and *P. lanceolata*

There are a number of species of plantain, all perennial, originally native to the countries of Europe and Central Asia, then spreading all over the world. The two best-known varieties in France have striated leaves. The greater plantain, *Plantago major*, or lamb's foot, has large leaves; while the ribwort plantain, *P. lanceolata*, or lamb's tongue, has narrow leaves. The two species have insignificant and drab flowers. In the past, the plantain ranked among the most popular of medicinal plants. A nettle or wasp sting, or a mosquito bite? A leaf rubbed on the affected area was enough to relieve the pain. A sore throat? A little of the bitter root flavoured with honey would cure it. A pimple on the nose? A cream made from plantain would get rid of it as quickly as it came. Conjunctivitis? A few leaves immersed in boiling water and placed over the eyes would relieve the ailment. Today, science has explained all those benefits. The plantain contains both mucilage, a natural softening agent, and tannins, astringent agents with an highly effective anti-inflammatory action. The plant is also made into an infusion for internal or external use to calm coughs and inflammation of the gums, to soothe haemorrhoids and ease constipation.

PLANTAIN

GENRE DES PLANTAGINÉES

PLANTAGO MAJOR

BLACK PEPPER
Piper nigrum

Contrary to popular opinion in France, the pepper plant, *poivrier*, was not discovered by Pierre Poivre, the famous adventurer at the service of the French East India Company, because it has been known since ancient times. Its name is derived from the Sanskrit *pilpali*, which is also the root for the words 'pepper' in English, *pepe* in Italian and *Pfeffer* in German. The pepper plant grows wild in India and Malaysia. From there it travelled by a series of convoluted routes to the West, where it reached exorbitant prices during the Middle Ages and was accepted as legal currency. The legend of Tristan and Isolde tells of two lovers who fell madly in love after drinking a love potion containing a pinch of pepper. Less poetically, tanneries would use pepper to repel insects and parasites, thanks to piperine, the pungent alkaloid it contains. From the eighteenth century onwards, the cultivation of pepper spread to all the humid and hot regions of the world. The pepper plant is a vine and requires a support, living or otherwise, to grow. Each flower produces between twenty and thirty berries. If they are picked before ripening, they are known as green peppercorns, and after drying in the sun, black peppercorns. White peppercorns are the inner seed with the outer layers removed. For medicinal use, pepper stimulates the digestion and gastric secretions. It is also effective for nausea and loss of appetite.

Le Poivre

PURSLANE
Portulaca oleracea

This small plant trails over the ground. Its thick red stalks bear thick, fleshy and shiny leaves and small yellow flowers that bloom throughout the summer. A native of Asia, purslane has spread around the world. It is commonly found in gardens, where it is often considered undesirable. However, this is a mistake, because it can be used for both culinary and medicinal purposes. Purslane is eaten in salads, dressed with olive oil, or boiled like spinach to bring out its deliciously tart flavour. It is a source vitamins C and E, fatty acids, omega-3 and beta-carotene. Consequently, it is a useful treatment for ageing, for the prevention of heart conditions and cardiovascular disease, and for protecting the arteries. Purslane is also high in mucilage, which gives it laxative and anti-inflammatory properties. As a poultice, it relaxes tensed muscles and can be used to treat inflammation of the eyelids. Its medical qualities are recognized around the world. It is used in China to treat diarrhoea, in Central America to treat gastritis, and in Africa as a sedative, among other uses.

POURPIER

COWSLIP
Primula officinalis

The cowslip, *Primula officinalis*, which grows in meadows and along paths, is a symbol of spring: it announces the arrival of the season. Its pale yellow petals form delicately scented bells that draw bees in search of nectar. It is a plant that prefers more northern climes, which explains why neither the Greeks nor the Romans mentioned it in their writings. In the Middle Ages, it gave rise to a number of stories and legends, superstitions and fables. Among them, it was believed to open the gates of heaven, leading to its common name Our Lady's keys and St Peter's wort. Hildegard of Bingen, a Benedictine abbess and scholar from the twelfth century, recommended it to drive away melancholy. It was also believed that cowslips could cure paralysis. Obviously, neither of these properties has been proven. Nevertheless, thanks to its anti-inflammatory properties and its ability to fight respiratory infections, cowslip has shown to be extremely effective. As an infusion, it acts as an expectorant for bronchitis and soothes coughs. As a compress, it is used to treat rheumatism and gout, sprains and other muscular pains. Its delicately flavoured flowers taken as an infusion calms restless infants.

PRIMULA OFFICINALIS
PRIMEVÈRE

PERUVIAN BARK
Cinchona officinalis

Spanish Jesuit priests discovered natives of the Amazonian Andes using Peruvian bark tree, also known as the cinchona tree, in the early seventeenth century. The Countess of Chinchón, wife of the Viceroy of Peru, was said to have been cured of her fevers by this plant. Whether or not this is true, the name of this Spanish aristocrat was given to the plant, in its Latin form. After its discovery, the ground bark of this tree was almost immediately sent back to Europe, where it was known by the names Jesuit's powder and the Countess's powder and used for the treatment of all sorts of fevers. However, it would not be until 1920 that its active ingredient, quinine, was isolated by two French pharmacists. Today, Peruvian bark is grown in tropical regions with heavy rainfall such as Hawaii, Indonesia and the Democratic Republic of Congo. There are several species of Peruvian bark, all with very similar properties. The active part of this large tree is found in its dried bark. Other than fevers, mainly malarial fever, it is effective for digestive tract problems and for lack of appetite owing to its astringency and bitterness. As a compress, it heals the skin and acts as a hair tonic.

Quinquina. (Cinchona Calisaya).

Décortication.

LIQUORICE
Glycyrrhiza glabra

This is a shrub from the south of Europe. Since the thirteenth century, it has been grown in gardens, but it is also found in scrubland and meadows of southern France, where it grows wild. Its strong rhizome develops runners, known as stolons, which are dried to produce the famous liquorice sticks that children love. In ancient times, the Greeks were already using this plant to quench their thirst and calm coughs. In the Middle Ages, liquorice was used to clear the voice, to treat respiratory ailments and for stomach pains. At the time, it was drunk as a refreshing tisane or the root was slowly chewed. In the late eighteenth century, liquorice became fashionable with cocoa, the famous drink sold by the glass in the streets of Paris or as a powder to dissolve in water. Medical uses for liquorice include as an expectorant and for the relief of coughs, for plaque removal from teeth, protection against tooth decay, and for the treatment of impeligo and eczema. It is also a valuable antidote for tobacco because it helps to renew the mucosa of the respiratory tract. However, care must be taken. If consumed in excess, liquorice can cause headaches and increase blood pressure.

RÉGLISSE
GENRE DES LÉGUMINEUSES

GLYCYRHIZA GLABRA

RHUBARB
Rheum rhabarbarum and *R. rhaponticum*

There are several different species of rhubarb, the best known of which are garden rhubarb, *Rheum rhabarbarum*, and false rhubarb, *R. rhaponticum*. All varieties have red stalks and large leaves, where, after it has rained, insects arrive to swim and butterflies to drink. Rhubarb is thought to have originated on the wind-swept steppes of Mongolia. The Chinese knew of it over four thousand years ago, but more for its therapeutic, vermifuge, tonic and purgative properties than for its culinary use. In around the eleventh century, the Arabs introduced it to Europe, but they only found the root to be of interest for the treatment of jaundice. The great Swiss physician Paracelsus confirmed this use in the sixteenth century. In the following century, rhubarb truly became a common plant in European gardens, where it was grown for culinary use. By this time it arrived from Russia, spreading onwards to England in the eighteenth century, where its tart stems would be used to make pies and chutneys. On the medicinal side, rhubarb can be used to treat occasional constipation, but it must be used with care because it can irritate the kidneys and cause inflammation of the bowel. Although it is delicious in a pie, it is best not to eat too much.

RHUBARBE
GENRE DES POLYGONÉES
RHÉUM

ROSEMARY
Rosmarinus officinalis

Commonly found in Mediterranean scrubland, rosemary is a small evergreen shrub with a pronounced flavour. It is susceptible to frost. Bees gather nectar and pollen from its bluish flowers, which they turn into a quality honey. The ancient Greeks and Romans held this plant to be a symbol of love, but they did not have any medicinal use. In the Middle Ages, essential oil was successfully extracted from the plant. During the Renaissance, the distillation of its flowers led to the production of the famous Hungary water, also known as Queen of Hungary's water. According to legend, this marvellous water was named in honour of Elizabeth of Poland, who after using a large amount of it, was cured of rheumatism overnight. Its effectiveness was so great that the lady, then over seventy years old, seduced the king of Hungary, who married her. Today, doctors acknowledge the restorative and invigorating powers of rosemary because it helps to boost the body's immune system. An infusion of flowering tops and leaves is recommended for people suffering from exhaustion. This beverage also aids digestion and is an expectorant. When rubbed on or used in a hot compress, rosemary has an anti-inflammatory action. It also heals the skin.

ROMARIN

SAFFRON CROCUS
Crocus sativus

This crocus with its ephemeral purple flowers does not exist in a wild state. It is the result of selective breeding and cultivation of a wild species, perhaps native to India, in about 3500 BC. Its long red stigmas, the parts of the pistil that receive pollen, were already worth their weight in gold during ancient times. The Phoenicians used saffron crocus, the Persians introduced it to lands of the Orient and the Arabs brought it to Spain in the Middle Ages. The Crusades also certainly had a part in its spread to Europe. While it was first grown at random, saffron crocus brought fortune to the farmers of the Quercy and Gâtinais regions of France at the end of the seventeenth century. The uses for saffron crocus have changed over time. In ancient times, it was prescribed as an aphrodisiac and a stimulant. Medieval physicians used it to treat toothache and the plague. Nowadays, a number of French saffron crocus farmers make a living from their precious harvest, mainly destined for the kitchen. Saffron subtly flavours and colours tajines and risottos, cream of mussel soup and orange marmalade. The plant also has one recognized medical property; it is a sedative. However, as it takes at least two hundred flowers to make 1 gram of saffron, it is easy to imagine that this remedy is expensive.

SAFRAN

GENRE DES IRIDÉES

CROCUS SATIVUS

SOAPWORT
Saponaria officinalis

In the summer, soapwort grows over slopes and along paths, displaying its pretty and fragrant pink flowers. Ancient peoples were already using it for its cleaning properties. When immersed in water, it produced a foam that would emulsify stubborn fat, cleaning fabrics and whitening delicate linen. For centuries, washerwomen made great use of it, by the side of streams or in wash houses. Quite a different use was made by women conscious of their beauty. They would rub the plant on their hair to make it shinier or put a few leaves or roots of the plant in their bath to purify their bodies. Soapwort, also known as the soap plant, truly deserves its name. It owes its properties to its saponin content, a molecule that dissolves fat in water. On the medicinal side, soapwort is known for its depurative properties. However, a large dose taken internally can paralyse the respiratory system and stop the heart. Animals are intuitively wary of it. So, it is better to use it externally, for rinsing fragile hair or treating eczema.

SAPONAIRE

GENRE DES
CARYOPHYLLÉES SILÉNÉES

SAPONARIA

SAGE
Salvia officinalis

Whether wild or cultivated, sage is aromatic. Its rubbed leaves release a strong odour that is both spicy and camphorated. The best-known species is garden sage, also known in France as 'thé de Grèce' or 'tea of Greece'. This native of western Asia grows wild on the dry hillsides around the Mediterranean. It has a woody stem and beautiful purplish-blue flowers, and it does not lose its leaves in winter. In the Middle Ages, it was one of the medicinal plants grown in the gardens of monarchs, nobles and monasteries. It was considered a panacea at the time, to treat all sorts of ailments. It was even used to ward off death! Its name is derived from the Latin *salvare*, 'to heal', a reminder of this belief. Physicians of the famous Medical School of Salerno coined the proverb: 'Why should a man die, when he can go to his garden for sage?' It is obvious that sage is not a miraculous remedy. However, its leaves, in particular, have healing properties. In an infusion, they are both a tonic and anti-perspirant, and they reduce fever and aid digestion. They also act on problems associated with menopause and menstruation. As a compress, they heal wounds, and as a mouthwash, they soothe mouth ulcers and inflammation of the gums. The list of benefits would not be complete without mention of its culinary role, as a seasoning for meat, for instance.

LES PLANTES UTILES

SAUGE OFFICINALIS

WHITE WILLOW
Salix alba

A willow tree is easily identified by its fragrant springtime flowers, known as catkins, its silver foliage and its rounded shape. However, it is very difficult to distinguish between the multitude of species, subspecies and hybrids. In general terms, the white willow tree grows close to rivers, marshes and drainage ditches. Interestingly, this tree, which loves damp regions, serves to treat the ailments that are endemic to them. By the sixteenth century, it was known that a few small branches could reduce fever and soothe rheumatic pain and headache, all the symptoms of a prolonged stay in swampy areas. In 1830, the French pharmacist Pierre Leroux confirmed the white willow tree's properties for reducing fever by isolating salicin. This molecule is the precursor for acetylsalicylic acid, the active ingredient of aspirin. Today, decoctions made from the bark are recommended for the relief of rheumatism and fever. Unlike the synthetically produced aspirin, white willow does not irritate the stomach. An infusion made from the white willow catkins also acts as a sedative and is effective for insomnia and anxiety.

Feuilles

Chaton
de Saule

Saule (Salix alba L.)

SENNA
Cassia senna and *C. angustifolia*

Alexandrian senna, *Cassia senna,* and Indian senna, *C. angustifolia,* are two very similar species of small, bushy shrubs. The former is cultivated in the Nile Valley and the latter in the Tamil Nadu (formerly Madras) region in India, which is also the origin of another of its names, Tinnevelly senna, after a city (modern-day Tirunelveli) in the same region. It has yellow flowers, dull green leaves and fruit resembling flat pods containing light brown seeds. The recipes for preparing senna mainly have a medicinal use. In the Koran, the prophet Mohammed recommended senna and honey 'for all illnesses except death'. However, the Arabs mainly used the shrub as a purgative. They also believed that senna could be used to treat all excessive behaviours, including madness. Today, when prepared in a pharmacy, its fruit and bitter-tasting dried leaves can be used to treat occasional constipation caused by a change in diet or travel, or persistent constipation when a fibre-rich diet is ineffective. A word of warning: senna-based remedies should never to be prescribed to young children or pregnant women, nor should they to be used for prolonged treatments, because the plant can be excessively purgative.

SENÉ

GENRE DES CÉSALPINIÉES

SENE

CREEPING THYME
Thymus serpyllum

Despite the words of *Mon Petit Lapin* ('My Little Rabbit') and other French nursery rhymes, rabbits do not like creeping thyme. They hate its odour. A close relative of garden or common thyme, *Thymus vulgaris*, this tiny plant creeps over slopes, sandy soils and rocks exposed to the sun. It is recognized by its pink flowers tinged with dark purple. The gentle fragrance released by the plant comes from the leaves, which are covered in glands filled with aromatic oil. It can vary depending on the place where the plant grows: the fragrance of a creeping thyme growing on a dune by the sea will be weaker than that of a plant growing on a rock under the sun. The leaves and flowering tops have a special place in folk medicine. Taken as an infusion, they ease swelling and flatulence. The same infusion with honey soothes the throat, calms coughs and is a cure for a hangover. In a compress, the leaves and flowers relieve the pain of sciatica and sprains, and itching. They invigorate the body in a hot bath, and are a treatment for pharyngitis when used in a mouthwash. Creeping thyme is, of course, also used as a herb, an excellent seasoning for vegetable and meat dishes. However, the preference in this case is for a cultivated variety, *T. citriodorus*, known as lemon thyme, which is the most commonly used variety for cooking.

LES PLANTES UTILES

SERPOLET

COMMON MARIGOLD
Calendula officinalis

The common marigold is easy to identify by the orange colour of its flowers. It is slightly viscous to the touch, and some people find its particular smell unpleasant. The flowers are sensitive to the humidity in the air. When humidity is high, they ferment, announcing the arrival of rain. They also act as natural clocks, opening with the first rays of sunlight and closing as the light fades. It is uncertain whether the marigold was used in ancient times. However, it is known that medieval physicians used it to treat ailments of the digestive system and the eyes, and as an antidote for snakebite and hornet stings. In the Renaissance, marigold was recommended for the treatment of jaundice for the simple reason that its flowers were the same colour as bile. Today, the benefits of the flower heads have been recognized. They have antimicrobial, anti-inflammatory, antiseptic, antispasmodic and skin-healing properties. Compresses and poultices made from the petals of freshly cut flowers are used to treat burns and chilblains, sunburn and nappy rash. The red oil, obtained by macerating the flowers, smooths dry skin. Some people will prefer to buy calendula, made from marigolds, for the same effect. On the culinary side, salads can be enlivened with the orange petals or the young buds pickled in vinegar like capers.

LES PLANTES UTILES

SOUCI

MEADOWSWEET
Filipendula ulmaria

In the month of June, meadowsweet covers the wet meadows with its elegant creamy white flowers. The tall plant with beautiful red stalks is unmistakable, and it well deserves its vernacular name queen of the meadow. Rubbing the flowers and leaves releases a strong, rather pleasant smell of bitter almonds. But as they are harmless, they are used to flavour desserts and drinks, and are used to make herbal tea. Despite all its benefits, it took a long time for the medicinal properties of meadowsweet to be recognized. Unknown in ancient times and practically ignored in the Middle Ages, the plant was not studied until the mid-nineteenth century. It was then discovered to have diuretic, tonic and vulnerary properties. At that time, when everybody was obsessed with science, the parish priest of the village of Trémilly pointed out that meadowsweet was highly 'useful for women going through the change', without explaining why. However, meadowsweet is mainly renowned for treating the fever people were subjected to in swampy areas, precisely where it grows in total freedom. Science has since explained this quality by showing that the plant contains acetylsalicylic acid, the active ingredient of aspirin. Today, it is known that meadowsweet is an anti-inflammatory. An infusion of the flowery tops is recommended to relieve lower back pain and stiff neck. It is also diuretic and depurative, making it ideal for people suffering from gout.

SPIRÉE

GENRE DES ROSACÉES

SPIRÆA ULMARIA

COMMON ELDER
Sambucus nigra

The common elder is partial to nitrogen-rich soils, which is why it is common near homes and on empty plots of land. In mid-spring, this shrub burgeons with masses of fragrant flowers, which draw insects to collect its nectar and pollen. At the end of summer, it produces dark and juicy fruit, which is ideal for making jams and elderberry robs (these are a type of thick syrup). Birds adore them. The Greeks and Romans, followed by the physicians of the Middle Ages, saw the elder as providing a remarkable antidote to viper bite, gout and jaundice. But none of these properties have been proven by science. It is known, however, that the decoction of an infusion made with dried elderflowers is diuretic, depurative and anti-inflammatory. The flowers also provoke sweating, and are very effective against illnesses that cause rashes such as measles. The black elderberries, in turn, are high in vitamins A and C, and are both invigorating and cause a laxative effect. However, care should be taken, because they stain terribly.

SUREAU
GENRE DES
CAPRIFOLIACÉES SAMBUCÉES

SAMBUCUS NIGRA

TANSY
Tanacetum vulgare

The tansy is typically found in abandoned places and along railway lines and paths. It is also grown in gardens for its pretty, golden buttons, which bloom in the middle of the summer. The ancient Greeks and Romans seem to have absolutely no knowledge of it. Because it was native to eastern Europe, it would have been introduced to the western part of the continent during the great Barbarian invasions, in about the fourth century AD. One particular name it received in France was 'herbe aux vers' or 'worm grass', a name earned by its highly vermifuge properties. The seeds, which were once sold in France under the name of 'barbotine', were in fact used to eliminate pinworms, tapeworms and roundworms. The fresh leaves give off a strong smell, due to the presence of a toxic compound they contain, thujone. Because they are equally dangerous for humans, they are only used as natural insect repellents against fleas, mites and other parasites.

TANAISIE

GENRE DES COMPOSÉES

TANACETUM
VULGARE

LARGE-LEAVED LIME
Tilia platyphyllos

Together with its cousin, the small-leaved lime, *Tilia cordata*, this tree, which is also known as the linden, adorns the avenues and squares of towns and cities. These tall trees can live for up to a thousand years. The flowers are given many uses. They are picked on 24 June, midsummer or the feast of St John the Baptist, as soon as they bloom. Gently infused in a tisane, they are used to treat all sorts of ailment, such as indigestion, a heavy stomach, colds and flu, insomnia, nervousness and migraines. They also make a wonderful remedy for physical and mental fatigue. Used externally, the flower heads treat chapping and cracking of the skin and insect bites. Used in a steam bath or in a compress, they cleanse the skin and enhance the beauty of the face. Lime sapwood, the wood immediately behind the bark, is also beneficial. Ground into a powder, it makes an excellent diuretic for treating renal colic and rheumatism. Because it has no side effects, the lime is one of the plants humans can trust most.

TILLEUL
GENRE DES TILIACÉES

TILIA

COLTSFOOT
Tussilago farfara

The ephemeral yellow flowers of this plant found along the banks of large watercourses open in early February. Later comes the turn of the large, thick, and velvety heart-shaped leaves. Because the flowers precede the leaves, coltsfoot was once called *filius ante patrem*, meaning 'the son before the father'. Its scientific name *Tussilago* comprises two Latin words *tussis*, 'cough', and *agere*, 'dispel', which is a reminder that the flowers and leaves of this plant have been used in infusions since ancient times to treat coughs, bronchitis and colds. Coltsfoot is one of the so-called 'pectoral plants', which can be used to make an excellent home-made syrup for treating ailments of the chest. The flowers also have sudorific and depurative properties. The leaves, in turn, make a perfect tobacco substitute without the harmful effects. What is more, coltsfoot helps to restore the mucosa of the respiratory tract.

TUSSILAGE

GENRE DES COMPOSÉES

TUSSIS

COMMON VALERIAN
Valeriana officinalis

This tall and elegant plant has white flowers with a delicate pink tinge and a striated green stalk. It is found almost everywhere in Europe, except the Mediterranean region, and grows beside ditches and along the banks of streams. It is recognized by its odour, especially when the flowers drop off. Cats love this plant because it produces a state of euphoria in them, hence one of its common names, cat's valerian. It is also known as cut-heal, and in France as 'herbe à la femme battue' or 'beaten wife's grass', because it was once used to relieve the pain of wounds and bruises. But it is mostly used to calm nervous complaints. It has actually been known since Renaissance times that the active ingredients contained by the root can be used to calm people suffering from irritation and migraines, to reduce fever, prevent epilepsy and asthma attacks, palpitations, relieve rheumatic pains and anxiety, and lower blood pressure, among others. However, it is renowned above everything else for its use as a sedative.

VALÉRIANE

GENRE DES VALÉRIANACÉES

VALERIANA OFFICINALIS

COMMON VERVAIN
Verbena officinalis

Common vervain, also known as common verbena, is native to Europe. It is a spindly plant with thin mauve flowers and a stiff stalk that grows very quickly along river banks and paths. It would have gone unnoticed had the ancient Greeks and Romans not revered it to the point of considering it a sacred herb. They would weave wreaths of the plant to crown the heads of ambassadors. It was also believed to immediately heal the wound caused by a spear. In the Middle Ages, it was attributed with a number of magical powers: it allowed the future to be predicted and offered protection from storms, ghosts and bad luck. It was also considered an aphrodisiac, ideal for rekindling the flame of dying love. This aromatic and bitter plant was later studied by the science of medicine. Its fresh leaves and stalks, in an infusion, were found to have anti-inflammatory properties and to strengthen the immune system. They would be used to treat nervous fatigue, throat problems, calm migraines and neuralgia, and encourage digestion and the secretion of bile. Used externally in the form of a poultice made from the boiled plant, vervain soothes chapped and cracked skin, and can treat sinusitis and rheumatism.

VERVEINE
GENRE DES
VERBÉNACÉES VERBÉNÉES
VERBENA

GRAPE VINE
Vitis vinifera

The grape vine has been cultivated since at least 2000 BC. It has always been a part of folk medicine. In the sixteenth century, Matthiole earnestly celebrated wine, which 'purges the brain, excites the intellect, delights the heart, lifts the spirits, cleanses the blood and expels all impurities from the body'. Nevertheless, the Italian physician and botanist had already acknowledged that it could also cause terrible ailments. Today, the polyphenols, plant-based tannins, contained in grape vines and wine are believed to provide protection against cardiovascular diseases. However, there is still a long list of medical properties attributed to the plant: the leaves are tonic and astringent; fresh grape juice is high in vitamins and minerals; the pomace is a stimulant; and grapeseed oil can relieve diarrhoea. More particularly, the leaves of the red climbing vine, a variety of grape vine with red leaves, contain active ingredients that protect the blood vessels. A decoction made from the leaves also relieves tired legs, venous insufficiency, haemorrhoids and bruises.

LES PLANTES UTILES

LA
VIGNE

Name	Family	Species	Other names
Hemp	Cannabaceae	Cannabis sativa	
Herb Robert	Geraniaceae	Geranium roberticanum	
Juniper	Cupressaceae	Juniperus communis	
Large-leaved lime	Tiliaceae	Tilia platyphyllos	
Lemon balm	Lamiaceae	Melissa officinalis	
Lesser periwinkle	Apocynaceae	Vinca minor	Common periwinkle
Liquorice	Fabaceae	Glycyrrhiza glabra	
Maize	Poaceae	Zea mays	
Marsh mallow	Malvaceae	Althaea officinalis	
Meadowsweet	Rosaceae	Filipendula ulmaria	Queen of the meadow
Mint	Lamiaceae	Mentha pulegium, M. piperita and M. viridis	Flea mint
Myrtle	Myrtaceae	Myrtus communis	
Oat	Poaceae	Avena sativa	
Olive tree	Oleaceae	Olea europaea	
Opium poppy	Papaveraceae	Papaver somniferum var. album and P. somniferum var. nigrum	
Orange tree	Rutaceae	Citrus vulgaris	
Oregano	Lamiaceae	Origanum vulgare	
Passion flower	Passifloraceae	Passiflora incarnata	Apricot vine
Peppers	Solanaceae	Capsicum annuum and C. frutescens	
Peruvian bark	Rubiaceae	Cinchona officinalis	

Name	Family	Species	Other names
Plantain	Plantaginaceae	*Plantago major* and *P. lanceolata*	
Purslane	Portulacaceae	*Portulaca oleracea*	
Rhubarb	Polygonaceae	*Rheum rhabarbarum* and *R. rhaponticum*	
Roman camomile	Asteraceae	*Chamaemelum nobile*	
Rosemary	Lamiaceae	*Rosmarinus officinalis*	
Saffron crocus	Iridaceae	*Crocus sativus*	
Sage	Lamiaceae	*Salvia officinalis*	
Senna	Fabaceae	*Cassia senna* and *C. angustifolia*	
Soapwort	Caryophyllaceae	*Saponaria officinalis*	Soap plant
St John's wort	Hypericaceae	*Hypericum perforatum*	Devil chaser, cammock
Strawberry tree	Ericaceae	*Arbutus unedo*	
Tansy	Asteraceae	*Tanacetum vulgare*	Hind-heal
Tasmanian blue gum	Myrtaceae	*Eucalyptus globulus*	
White deadnettle	Lamiaceae	*Lamium album*	
White horehound	Lamiaceae	*Marrubium vulgare*	Hound's bane
White willow	Salicaceae	*Salix alba*	
Wild pansy	Violaceae	*Viola tricolor*	
Yarrow	Asteraceae	*Achillea millefolium*	Woundwort, devil's nettle

FURTHER READING

Collectif, *Phytothérapie, la santé par les plantes*, Sélection du Reader's Digest, Vidal, 2010.

Fournier (Paul), *Le Livre des plantes médicinales et vénéneuses de France. 1 500 espèces par le texte et par l'image*, Paul Lechevalier, 1947.

Girre (docteur Loïc), *Connaître et reconnaître les plantes médicinales*, Ouest France, 1980.

Gurib Fakim (Ameenah), *Toutes les plantes qui soignent : plantes d'hier, médicaments d'aujourd'hui*, Michel Lafon, 2008.

Künkele (Ute) et Lohmeyer (Till R.), *Plantes médicinales, identification, récolte, propriétés et emplois*, Parragon Books, 2007.

Leclerc (Henri), *Précis de phytothérapie*, Masson & Cie, 1932.

Lieutaghi (Pierre), *Le Livre des bonnes herbes*, Actes Sud, 1996.

Lieutaghi (Pierre), *Le Livre des arbres, arbustes et arbrisseaux*, Actes Sud, 2004.

Lieutaghi (Pierre), *Badasson & Cie. Tradition médicinale et autres usages des plantes en haute Provence*, Actes Sud, 2009.

Pamplona-Roger (docteur George), *Guide des plantes médicinales*, « Encyclopédie Vie et Santé », 1996.

Rombi (Max) et Robert (Dominique), *120 plantes médicinales : composition, mode d'action et intérêt thérapeutique… de l'ail à la vigne rouge*, Alpen, 2007.

IN THE SAME COLLECTION IN FRENCH

Little Book
of Saints

Little Book
of Saints Vol. II

Little Book
of Angels

Little Book
of Mary

Little Book
of the Bible

Little Book of
the Life of Jesus

Little Book
of the Gods

IN THE SAME COLLECTION IN FRENCH

Little Book of
Kings of France

Little Book
of Napoleon

Little Book of the
French Revolution

Little Book of
French Departments

Little Book of Towns
and Coats of Arms

Little Book of
Idiomatic Expressions

Little Book
of Castles

Little Book of Paris
(available in English)

Little Book
of Gourmet France

IN THE SAME COLLECTION IN FRENCH

Little Book
of Trees

Little Book
of Flowers

Little Book
of Mushrooms

Little Book
of Cats

Little Book of Dogs
(available in English)

Little Book
of Babies

For more information and follow our news:

- Please visit **www.editionsduchene.fr**
- Sign up for our newsletter on our website
- Follow Éditions du Chêne on

All images are from the private collection
of Papier Cadeau.

For the original edition:
© 2013 Papier Cadeau – Hachette Livre
www.papier-cadeau.fr

Editorial Manager: Nathalie Bailleux
with the collaboration of Franck Friès
Editorial Assistant: Fanny Martin and Audrey Gérard
Art Director: Sabine Houplain, with the
collaboration of Claire Mieyeville and Audrey Lorel
Proofreading: Myriam Blanc
For the current edition:
Cover by Charles Ameline
Layout and Photogravure: CGI

Translation © Papier Cadeau – Hachette Livre, 2019
English translation and proofreading by John Ripoll
and Theresa Bebbington for Cillero & de Motta

Published by Papier Cadeau
(58, Rue Jean Bleuzen, 92178 Vanves Cedex)
Printed by Toppan Leefung Printing in China
Printed in December 2020
ISBN 978-2-81231-981-5
75/6342/6-03